RiverStream Science

DO YOU REALLY WANT TO VISIT THE MOON

BY THOMAS K. ADAMSON

ILLUSTRATED BY DANIELE FABBRI

RiverStream Science
Great Reading • Real Learning

Amicus Illustrated hardcover edition is published by
Amicus. P.O. Box 1329, Mankato, MN 56002.
www.amicuspublishing.us

Copyright © 2014 Amicus. International copyright
reserved in all countries. No part of this book may
be reproduced in any form without written permission
from the publisher.

RiverStream Publishing reprinted with permission of
The Peterson Publishing Company.

Library of Congress Cataloging-in-Publication Data
Adamson, Thomas K., 1970–
Do you really want to visit the moon? / by Thomas K.
Adamson ; illustrated by Daniele Fabbri. — 1st ed.
p. cm. — (Do you really want to visit—?)
Audience: K-3.
Summary: "A child astronaut takes an imaginary trip
to the Moon, visits the sites from the Apollo missions,
and decides that Earth is a good home after all.
Includes solar system diagram, Moon vs. Earth fact
chart, and glossary"—Provided by publisher.
Includes bibliographical references.
ISBN 978-1-60753-197-5 (library binding) —
ISBN 978-1-60753-404-4 (ebook)
1. Moon—Juvenile literature. 2. Moon—Exploration—
Juvenile literature. I. Fabbri, Daniele, 1978– ill. II.
Title. III. Series: Do you really want to visit—?
QB582.A328 2014
523.3—dc23   2012025973

Editor: Rebecca Glaser
Designer: The Design Lab

Printed in the United States of America at
Corporate Graphics in North Mankato, Minnesota.

1 2 3 4 5 CG 16 15 14 13
RiverStream Publishing—Corporate Graphics,
Mankato, MN—022014—1047CGW14

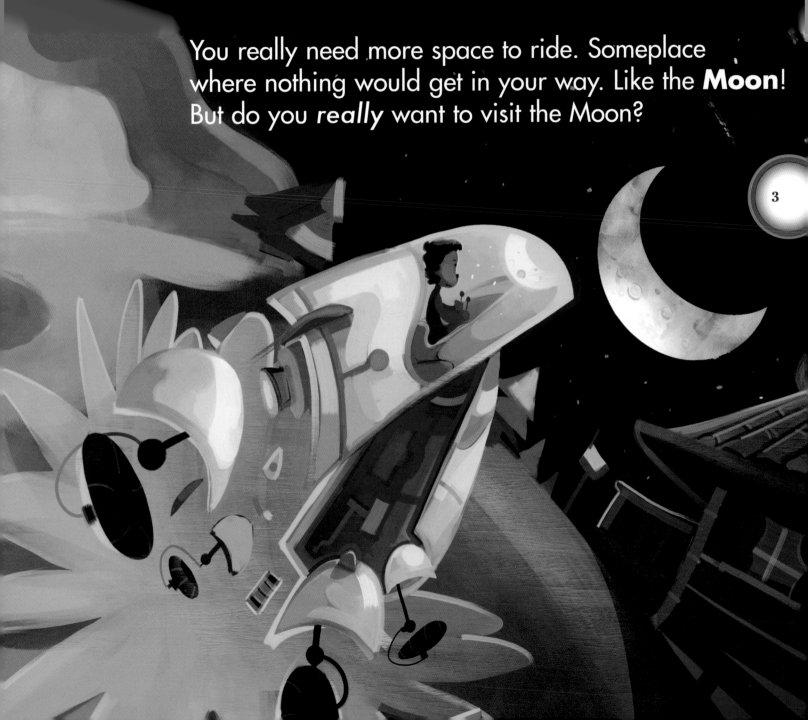

You really need more space to ride. Someplace where nothing would get in your way. Like the **Moon**! But do you *really* want to visit the Moon?

3

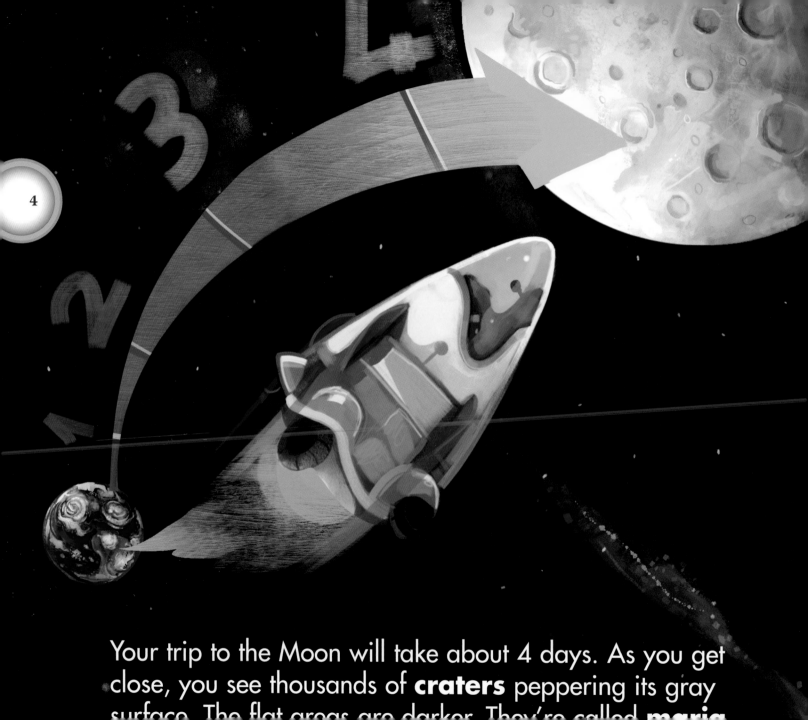

Your trip to the Moon will take about 4 days. As you get close, you see thousands of **craters** peppering its gray surface. The flat areas are darker. They're called **maria**

Some maria look nice and flat. They still have a lot of craters. And watch out for those **boulders**. They were kicked out of the ground when **meteorites** struck the Moon.

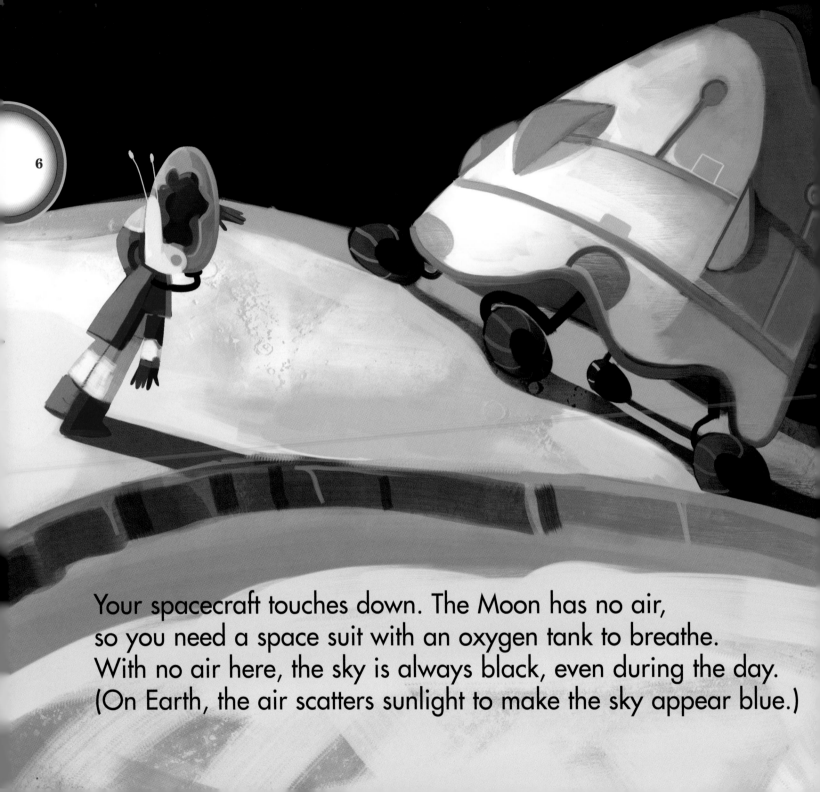

Your spacecraft touches down. The Moon has no air,
so you need a space suit with an oxygen tank to breathe.
With no air here, the sky is always black, even during the day.
(On Earth, the air scatters sunlight to make the sky appear blue.)

Walking on the Moon is like trying to walk in a sand dune. It takes a while to get used to. Hopping like a kangaroo seems to work well.

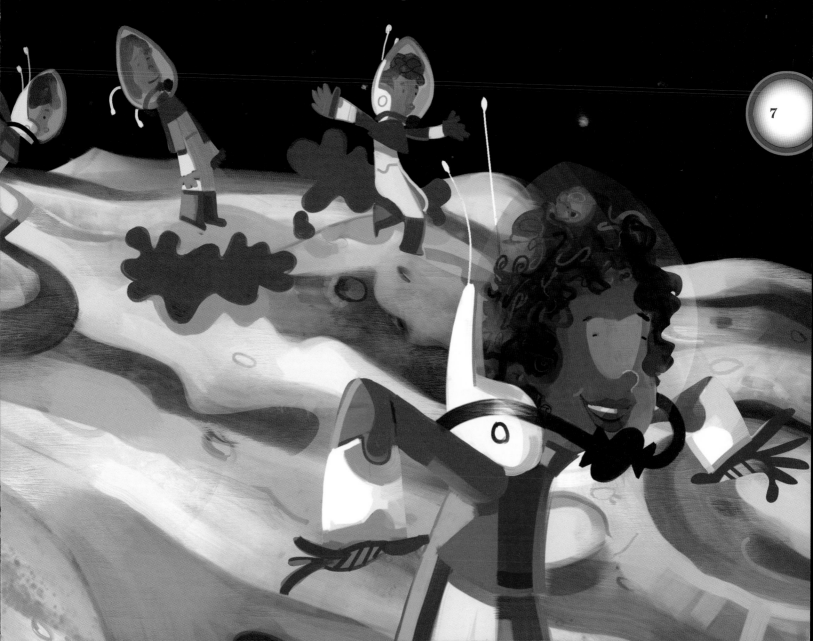

The Moon has less **gravity** than Earth.
If you weigh 60 pounds (27 kg) on Earth, you
weigh only 10 pounds (4.5 kg) on the Moon.
Slam dunks are really easy.

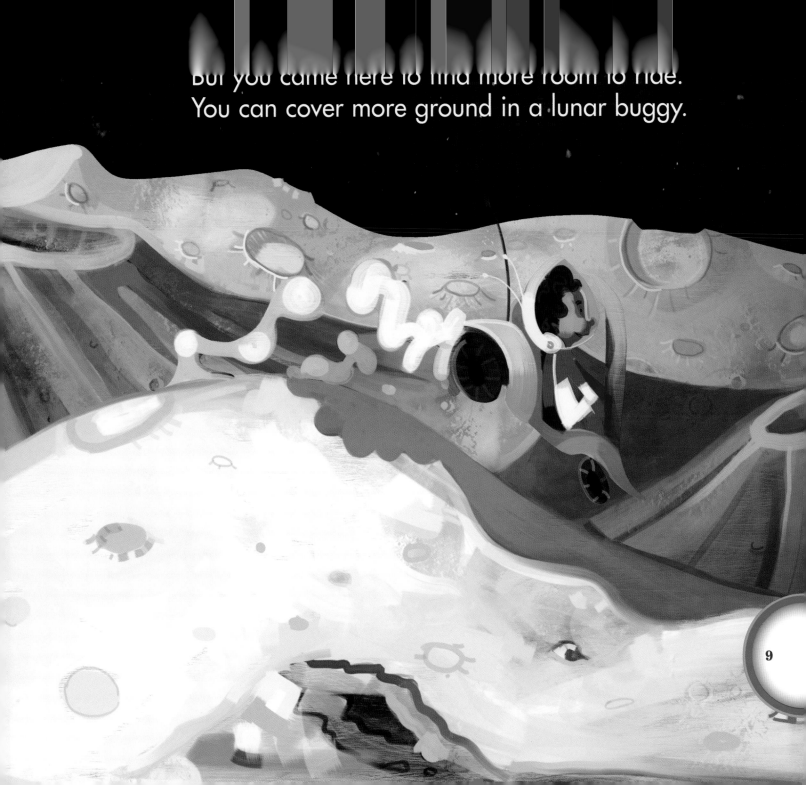

But you came here to find more room to ride.
You can cover more ground in a lunar buggy.

9

Wow! This is where the astronauts first walked on the Moon.

You can see their footprints. They're still here
more than 40 years later, because the Moon
has no wind or rain to wear them away.

11

Your space suit is getting very dirty. Moon dust sticks to everything. The dirt on the Moon's surface, like a fine powder, is called **regolith**.

Weren't there other astronauts on the moon? You cruise around and find five other landing sites.

You can't hear the electric motor of your **rover**. Hmm.
You try drumming on the dashboard. No sound.
Sound needs to travel through air to be heard. With
no air, there's no sound on the Moon. No music either.

14

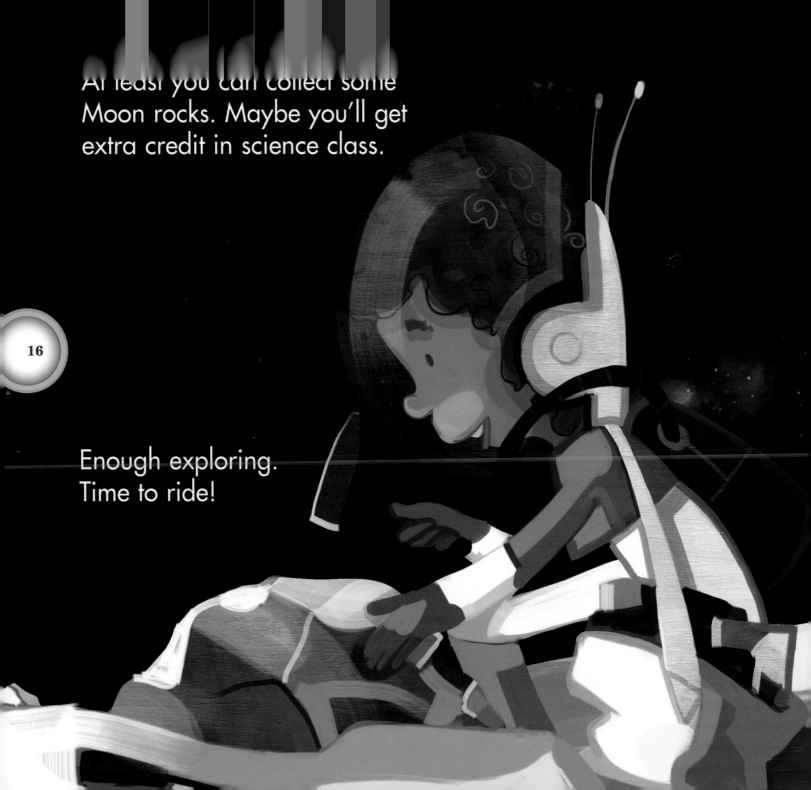

At least you can collect some Moon rocks. Maybe you'll get extra credit in science class.

Enough exploring.
Time to ride!

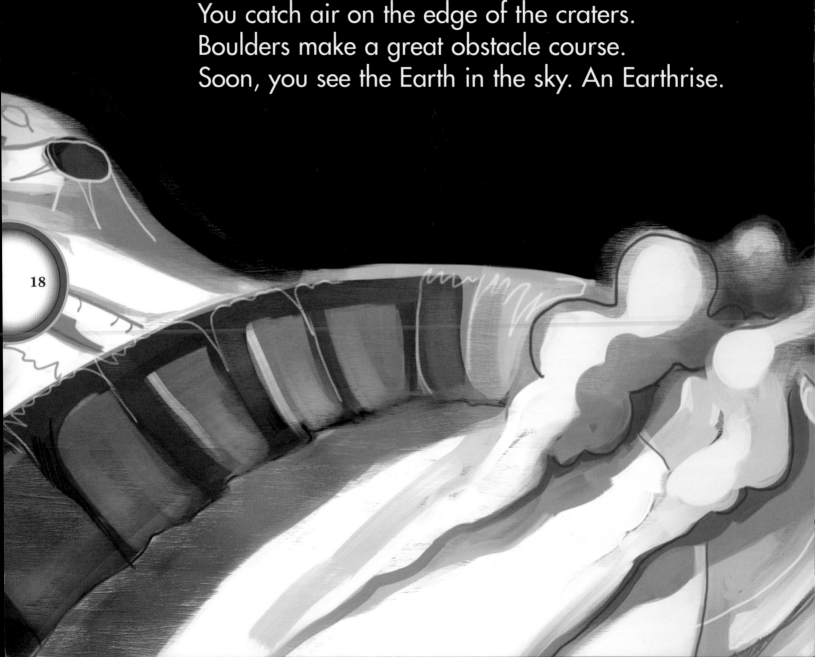

You catch air on the edge of the craters.
Boulders make a great obstacle course.
Soon, you see the Earth in the sky. An Earthrise.

18

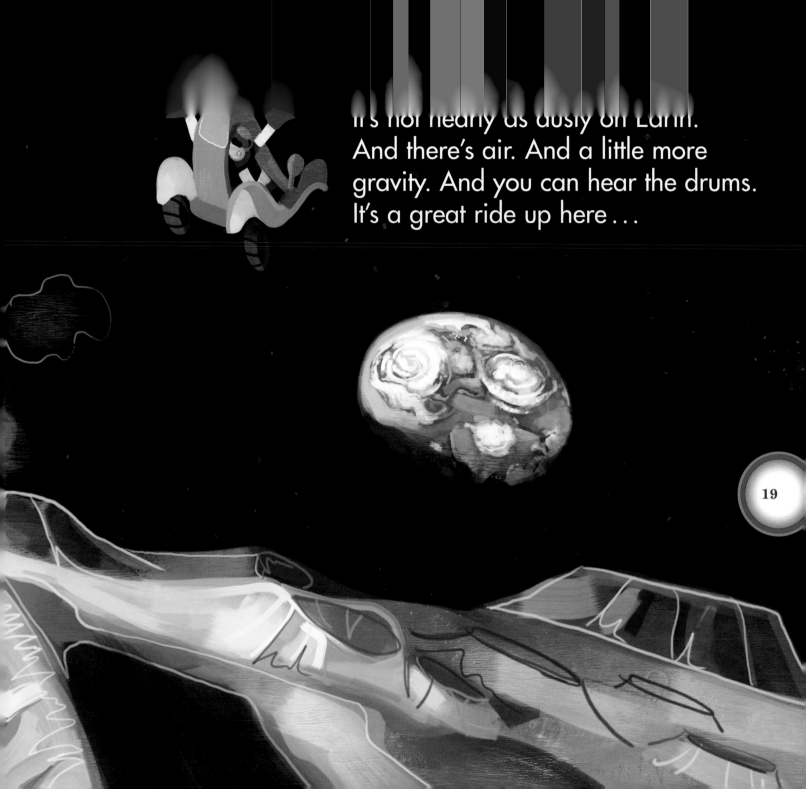

It's not nearly as dusty on Earth.
And there's air. And a little more
gravity. And you can hear the drums.
It's a great ride up here . . .

19

... but you wouldn't really want to stay on the Moon too long.

NEPTUNE

SATURN

URANUS

MOON

MARS

JUPITER

EARTH

VENUS

MERCURY

SUN

How Do We Know About the Moon?

We know about the Moon because astronauts have been there. Between 1969 and 1972, 12 astronauts walked on the Moon during the Apollo Project. Scientists are still studying the rock and soil samples they brought back to Earth. In 2009, the *Lunar Reconnaissance Orbiter* (LRO) went into orbit around the Moon. It is creating detailed maps of the surface.

Earth vs. Moon

	Earth	Moon
Position in solar system	Third from Sun	Orbiting around Earth
Average distance:	From Sun: 93 million miles (150 million km)	From Earth: 238,855 miles (384,400 km)
Orbit	365 days (around the Sun)	27.32 days (around Earth)
Day (sunrise to sunrise)	24 hours	29.53 days
Diameter	7,926 miles (12, 756 km)	2,159 miles (3,474 km)
Mass	1	About 81 Moons would equal the mass of Earth
Air	Oxygen and nitrogen	None
Water	About 70% covered with water	Water ice may exist in shadowed polar craters
Moons	1	0
Drummers	Need earplugs	No earplugs necessary, can't hear a thing.

Glossary

boulder A large, rounded rock.

crater A large hole in the ground caused by a piece of rock from space hitting a planet or moon.

gravity The force that pulls things down toward a planet or moon.

maria Large, flat plains on the Moon that appear dark from Earth.

meteorite A piece of rock or metal from space that hits a planet or moon.

moon A body that circles around a planet.

regolith A fine, power-like dirt that covers the surface of a moon or planet.

rover A vehicle for exploring the surface of a planet or moon.

Read More

Baker, David, and Heather Kissock. *Living on the Moon: Exploring Space*. Weigl, 2010.

Burleigh, Robert. *One Giant Leap*. Philomel Books, 2009

Floca, Brian. *Moonshot: The Flight of Apollo 11*. Atheneum Books for Young Readers, 2009.

Grego, Peter. *Exploring the Moon*. QEB Publishing, 2007.

Mist, Rosalind. *Sun and Moon*. QEB Publishing, 2013.

Websites

Explore the Apollo 11 Landing Site
http://www.nasa.gov/externalflash/apollo11_landing/
Interactive photographs allow users to see what it was like for the first astronauts on the Moon and inside the lander, and you can listen to transcripts of the landing.

Moon Exploration Pictures—National Geographic Kids
http://kids.nationalgeographic.com/kids/photos/moon-exploration/
View images of the astronauts and the moon's surface from the Apollo missions to the Moon.

NASA—Lunar Reconnaissance Orbiter
http://www.nasa.gov/mission_pages/LRO/main/index.html
View the newest images of the Moon's surface, sent by NASA's satellite in orbit around the Moon.

About the Author

Thomas K. Adamson lives in Sioux Falls, SD, with his wife, two sons, and a dog. He has written dozens of nonfiction books for kids, many of them about planets and space. He enjoys sports, card games, reading with his sons, and pointing things out to them in the night sky.

About the Illustrator

Daniele Fabbri was born in Ravenna, Italy, in 1978. He graduated from Istituto Europeo di Design in Milan, Italy, and started his career as a cartoon animator, storyboarder, and background designer for animated series. He has worked as a freelance illustrator since 2003, collaborating with international publishers and advertising agencies.